I0000301

K-Math Workbook Grade 7

A Smart Way of Learning Math

Jaehwa Choi, Ph.D.
The George Washington University

Sunhee Kim, Ph.D.
Howard Community College

Kyongil Yoon, Ph.D.
Notre Dame of Maryland University

CAFA Lab

How to Use K-Math Workbook

All Practice Pages are fully organized by the Common Core State Standards (CCSS). By scanning each **Item's QR code** with any smart device, one can instantly access these additional features:

Detailed Solutions
Get the answer with detailed solution steps

Item Drills
Drill the item variants until fully mastering the item

More Practice
Practice other items until fully mastering each standard

By scanning the **Exam QR code** in the Exam Page, one can submit answers and get immediate feedback such as: grades, correct answers, solution steps, additional practices, and diagnosis reports.

K-Math Workbook, integrating Information and Communication Technologies (ICT) into a classical paper workbook, is designed to cope with these Common Core State Standards challenges:

- Help to articulate to parents, teachers, and the general public expectations using a simple but Concrete Application
- Provide aligned textbooks, digital media and curricula to international standards with a Workbook Solutions
- Help to implement an assessment system to measure student performance on standards via a Formative Assessment System
- Provide professional development opportunities for educators on identified needs and more efficient practices with a Teacher Friendly Application
- Help evaluate policy changes needed to help students and educators, so they can meet the standards through a Research Oriented Approach

For more information about K-Math Workbook visit http://K-Math.CAFALab.com.

© 2016 by CAFA Lab, Inc.

CAFA Lab, Inc. All rights reserved. Printed in the United States of America. This publication is protected by Copyright and permission should be obtained from the publisher prior to any prohibited reproduction, storage in a retrieval system, or transmission in any form or by any means, electronic, mechanical, photocopying, recoding, or likewise. For information regarding permission(s), email to: admin@CAFALab.com. The website, www.CAFALab.com, information through QR code from this publication is protected by Copyright. You may not transfer or distribute the website contents to anyone else by any means, including by posting it on the Internet. K-Math Workbook is a trademark of CAFA Lab, Inc.

CAFA is a trademark of CAFA Lab, Inc.

Table of Contents

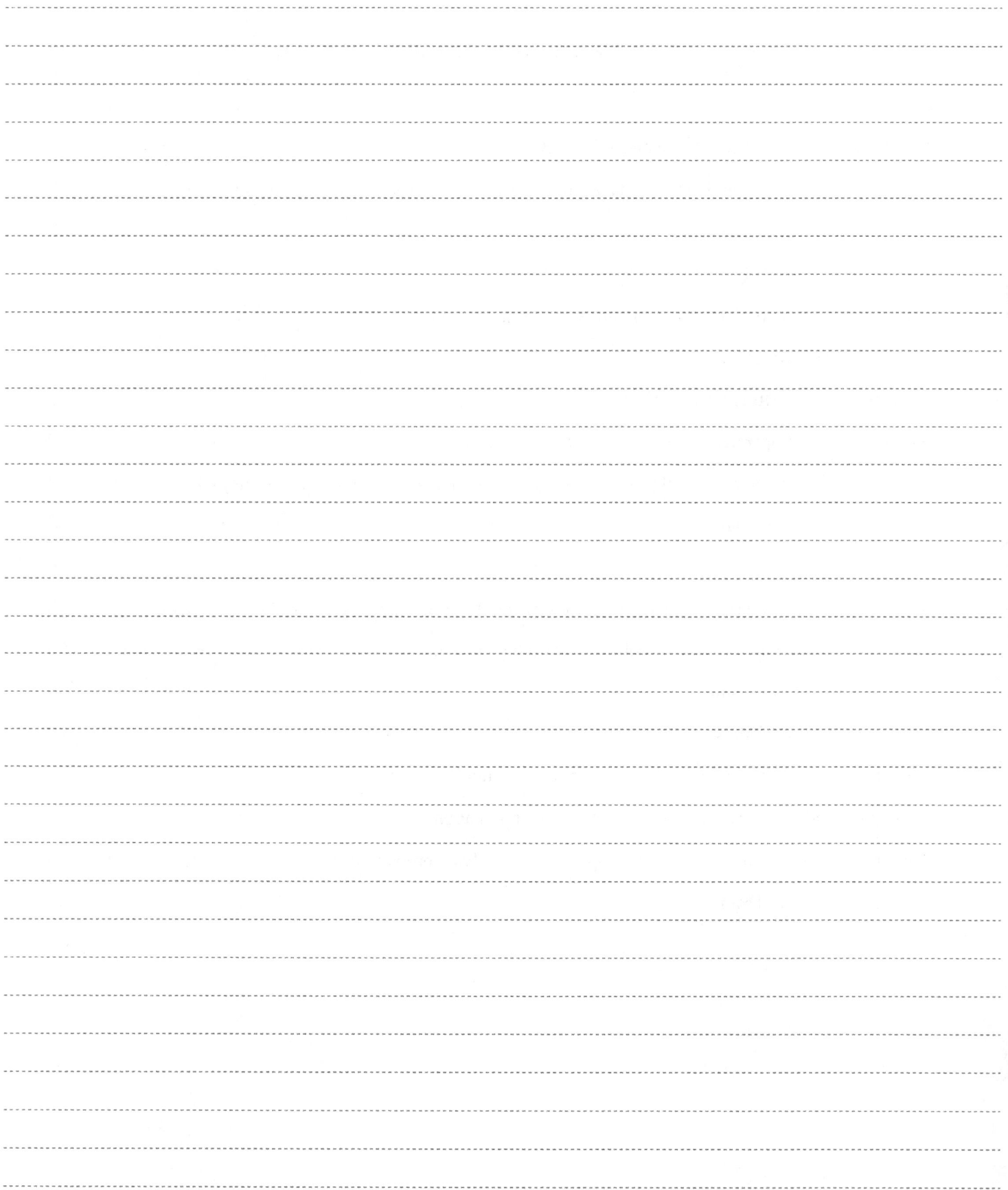

Ratios and Proportional Relationships

7.RP

1. Which of the following is equivalent to the ratio 4 : 4.2 ?

   Ⓐ 21 : 20   Ⓑ 20 : 21   Ⓒ 42 : 4

   Ⓓ 4.2 : 4   Ⓔ 4 : 42

2. If a person walks $\frac{1}{3}$ of a mile in $\frac{1}{6}$ of an hour, how many miles can he walk an hour?

3. If Megan can solve 5 math questions in 20 minutes, how many questions can she solve in an hour?

4. If Emily needs $\frac{1}{2}$ of a can of paint to paint $\frac{1}{14}$ of her room, how many cans of paint would she need to paint her room?

1. Which of the following represents a proportional relationship between the quantities below?

5	6
a	b

Ⓐ $5a = 6b$

Ⓑ $\dfrac{5}{a} = \dfrac{b}{6}$

Ⓒ $6 - 5 = b - a$

Ⓓ $5 + a = 6 + b$

Ⓔ $\dfrac{5}{a} = \dfrac{6}{b}$

2. Emily read the same number of pages of a book every day. If she read **63** pages in **9** days, what equation represents the number of pages that she read(P) related to the number of days(d) ?

3. Ethan drove to his friend's house, and the following graph shows the proportional relationship between the number of hours, x, and the distance in miles, y.

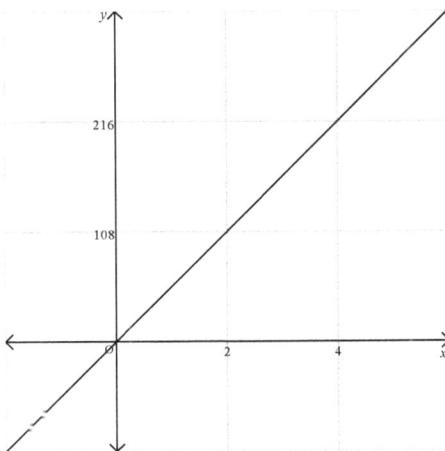

If it took him **5.5** hours to arrive, what is the distance to his friend's house in miles?

4. Ethan wanted to enlarge a photo that is **7** inches wide and **12** inches long to make a large poster. If the sizes of the photo and the poster are proportional and the width of the poster is **14** inches, what is the length of the poster?

1. Which of the following is proportional?

Ⓐ 5 to 3 and 12 to 20   Ⓑ 5 to 3 and 20 to 12

Ⓒ 5 to 3 and 5 to 7   Ⓓ 5 to 3 and 3 to 5

Ⓔ 5 to 3 and 7 to 5

2. Which of the following shows a proportional relationship between two quantities?

Ⓐ 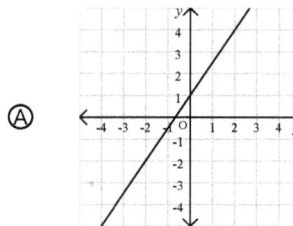   Ⓑ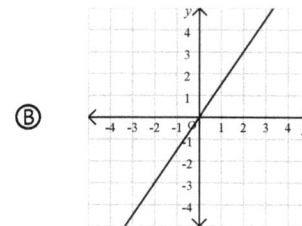

Ⓒ 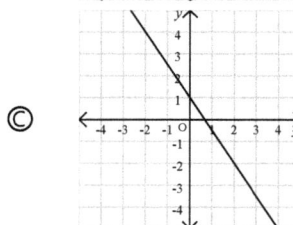   Ⓓ

4	5
16	24

Ⓔ

2	3
6	11

3. What value of a makes the following proportional?

7	14
4	a

4. Which of the following does not form a proportion with "**3 cups of sugar for 10 cakes**"?

Ⓐ 10 cups of sugar for **3** cakes

Ⓑ 6 cups of sugar for **20** cakes

Ⓒ 1.5 cups of sugar for **5** cakes

Ⓓ 10.5 cups of sugar for **35** cakes

Ⓔ 5.4 cups of sugar for **18** cakes

1. What is the constant of proportionality related to the number of pages in a book Simon has read per day?

Number of pages	28	36
Number of days	7	9

2. What is the constant of proportionality in $y = 9x$?

3. Christine paid $3.78 for 2 apples. What is the unit price of an apple?

4. The cost to rent a birthday party location is $135 for 5 kids, and $270 for 10 kids. What is the cost for 8 kids?

1. Which of the following represents a proportional relationship between x and y ?

   Ⓐ $y = 5x + 1$ 　　　　 Ⓑ $y = 2$

   Ⓒ $x = \dfrac{-5}{y}$ 　　　　 Ⓓ $xy = 25$

   Ⓔ $\dfrac{x}{y} = 5$

2. If Simon walks at a steady speed of 3 mph, what is the equation that represents the proportional relationship between the distance, d, and the time spent walking, t?

3. If total cost C is proportional to n items purchased at a price of \$11 each, what is the equation that represents the relationship between the total cost and the number of items?

4. y varies directly as x, and $y = 16$ when $x = 8$. What is y if $x = 20$?

1. What is the unit rate of the following graph?

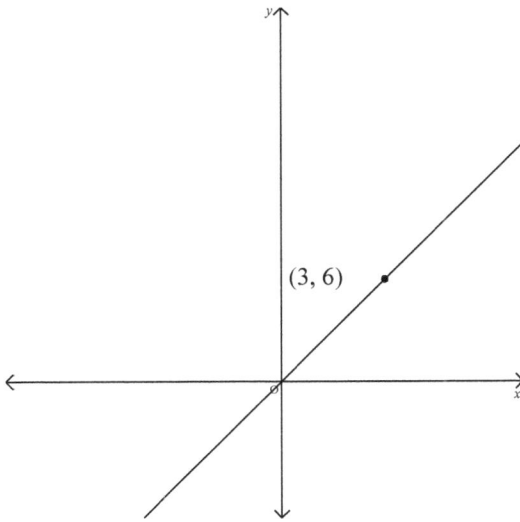

(3, 6)

2. Matthew drove with a steady speed of 60 mph. Which of the following shows the proportional relationship between the distance and the time where y represents the distance and x represents time(hours)?

Ⓐ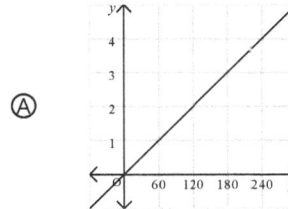

Ⓑ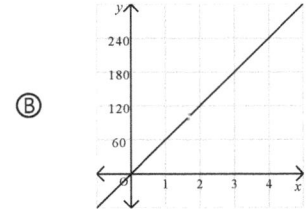

Ⓒ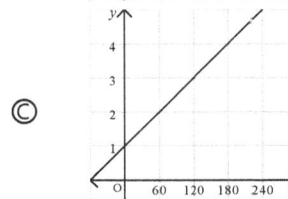

Ⓓ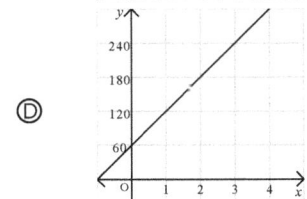

Ⓔ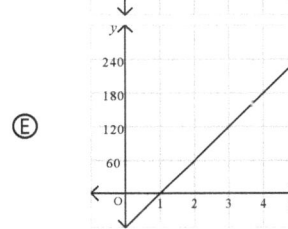

3. The following graph shows the proportional relationship between the distance, y, and the time, x. How many hours does it take James to arrive his destination if he is 306 miles away?

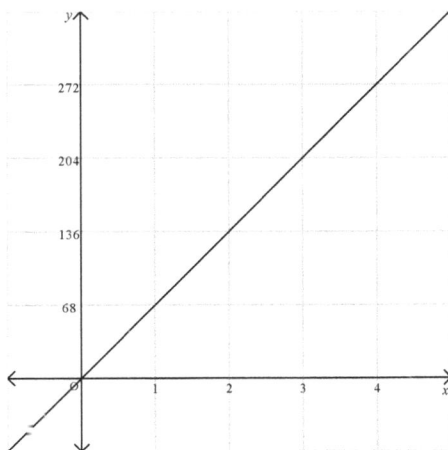

4. Which of the following gives the greatest unit rate when each point is on the graph of a proportional relationship?

Ⓐ (8, 80)　　　　Ⓑ (10, 1)

Ⓒ (15, 2)　　　　Ⓓ (1.5, 13.5)

Ⓔ (2, 30)

1. Last year, the price of 9 pies was $180. This year, the price of 3 pies is $66. What is the percent increase of one pie?

2. What equation represents "19% of a number is 94"?

   Ⓐ $94 = 19 \bullet n$ Ⓑ $n = 19 \bullet 94$

   Ⓒ $n = 0.19 \bullet 94$ Ⓓ $19 = 94 \bullet n$

   Ⓔ $94 = 0.19 \bullet n$

3. The population of a city is about 13% of the entire county's population. If the population of the county is 60000 people, what is the population of the city?

4. The regular price of one dish is $22. If a 26% discount coupon is used, what is the price paid?

1. Fred paid $99.84 for 4 bags of 4 pound birdseed. What is the price per ounce?

Ⓐ $0.82 Ⓑ $0.52 Ⓒ $0.59

Ⓓ $0.29 Ⓔ $0.39

2. What equation represents the relationship between x and y?

x	15	18	24
y	5	6	8

Ⓐ $y = x + 3$ Ⓑ $y = x - 3$

Ⓒ $y = \dfrac{3}{x}$ Ⓓ $y = \dfrac{x}{3}$

Ⓔ $y = 3x$

3. A local stationery store sells one pen for 140 cents and a boxed set of 3 pens for 294 cents. If you buy 3 pens in a box, what is the discount rate, in percent, compare to buy 3 pens separately?

Ⓐ 30% Ⓑ 3% Ⓒ 0.7%

Ⓓ 70% Ⓔ 7%

4. Which of the following ordered pairs are in a proportional relationship?

Ⓐ $(4, 1), (12, 15)$ Ⓑ $(4, 1), (12, 3)$

Ⓒ $(4, 1), (3, 12)$ Ⓓ $(4, 1), (15, 3)$

Ⓔ $(1, 4), (12, 3)$

5. Emily paid $3.96 for 4 apples and $4.72 for 8 oranges. How much would she have to pay for one apple and one orange?

Ⓐ $1.98   Ⓑ $1.18   Ⓒ $1.58

Ⓓ $0.99   Ⓔ $0.59

6. y varies directly as x, and $y = 8$ when $x = 10$. What is x if $y = 16$?

Ⓐ 17   Ⓑ 24   Ⓒ 22

Ⓓ 20   Ⓔ 21

7. The triangles $\triangle ABC$ and $\triangle DEF$ are similar. The perimeter of the triangle $\triangle ABC$ is 170, and the lengths for the corresponding sides for $\triangle ABC$ and $\triangle DEF$ are 21 and 42, respectively. What is the perimeter of $\triangle DEF$?

Ⓐ 342   Ⓑ 233   Ⓒ 382

Ⓓ 212   Ⓔ 340

8. Megan had dinner with her friends in another state. They ordered a total $36 on the meals, and the bill was $37.62 including tax. What is the tax rate in that state?

Ⓐ 5.5%   Ⓑ 4%   Ⓒ 4.8%

Ⓓ 4.3%   Ⓔ 4.5%

9. Find the ratio which is equivalent to 4 : 4.4.

  Ⓐ 10 : 11   Ⓑ 4 : 44   Ⓒ 44 : 4

  Ⓓ 4.4 : 4   Ⓔ 11 : 10

10. Ave wanted to enlarge a photo that is 4 inches wide and 14 inches long to make a large poster. If the sizes of the photo and the poster are proportional and the width of the poster is 24 inches, what is the perimeter of a poster?

  Ⓐ 216   Ⓑ 48   Ⓒ 168

  Ⓓ 108   Ⓔ 240

11. When a graph passes through $(0, 0)$ and $(8, 2)$, what is the unit rate between x and y?

  Ⓐ 4   Ⓑ $\frac{1}{4}$ Ⓒ $\frac{1}{2}$

  Ⓓ 8   Ⓔ $\frac{1}{8}$

12. What percent of 80 is 60?

  Ⓐ 48%   Ⓑ 75%   Ⓒ 7.5%

  Ⓓ 0.75%   Ⓔ 4.8%

13. At Joanne's school, 45% are girls and 50% of them wear glasses. If the number of boys in her school is 330, how many girls wear glasses?

   Ⓐ 135   Ⓑ 25   Ⓒ 125

   Ⓓ 165   Ⓔ 300

14. At the zoo, the ratio of adult admission fee and child admission fee was $10 : 9$ before the price increase. After the admission fee for an adult and a child was increased by \$8, the admission fee for an adult becomes \$18. What is the admission fee for a child after the increase in price?

   Ⓐ \$16   Ⓑ \$15   Ⓒ \$18

   Ⓓ \$17   Ⓔ \$19

15. Last year, the total number of employees of one company was 420, and the number of male employees was $\frac{5}{9}$ of the number of female employees. This year, the number of male employees has increased by 30% and the number of female employees has decreased by 10%. What is the total number of employees this year?

   Ⓐ 615   Ⓑ 195   Ⓒ 438

   Ⓓ 423   Ⓔ 528

16. Annual interest rate is 8%. If the initial deposit is \$300, what is the accumulated amount at the end of 4 years? (The interest rate applies only on the initial deposit.)

   Ⓐ \$2400   Ⓑ \$300   Ⓒ \$312

   Ⓓ \$308   Ⓔ \$396

The Number System

7.NS

1. A submarine was **735** feet below sea level. If it ascends **90** feet, what is its new position?

   Ⓐ 90 feet below sea level

   Ⓑ 645 feet below sea level

   Ⓒ 825 feet below sea level

   Ⓓ At sea level

   Ⓔ 735 feet below sea level

2. What is $(-19) + (-10)$?

3. Matthew had a balance of **$30** in his bank account. Without considering his balance, he wrote checks for **$25.96** and **$5.68**. What will Matthew's account balance be after the two checks go through?

4. An owner of a shopping mall recorded the monthly profit in the first half a year. The profits of January, February, and April are **3.1** millions, **3** millions, and **7.8** millions, respectively. In March, May, and June, the shopping mall lost **5.2** millions, **5.4** millions, and **3.6** millions, respectively. What is the total profit for the first half a year?

1. What is opposite of 4?

2. Ave owes his friend 14 dollars. The next day, he pays his friend 11 dollars, and two days later, he pays him 3 dollars. How much money does Ave owe him after two days?

3. Find all the values of x that satisfy $|x| = 16$.

4. A helicopter was flying at the height of 5000 feet above ground. The helicopter first moved up 260 feet, then moved down 500 feet, and then moved up 240 feet. What is the height of the helicopter from the ground?

1. Which of the following is an expression of the number line below?

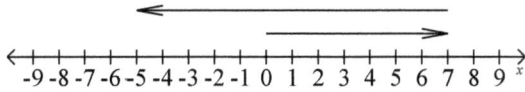

Ⓐ $-5 + 7$ Ⓑ $7 - 12$

Ⓒ $7 + 12$ Ⓓ $12 - 7$

Ⓔ $5 - 7$

2. What is the number that is located at a distance of 16 from the left of -13?

3. There are three houses in a row. James's house locates 25 feet to the right of Emily's house, and Joan's house is 16 feet to the left of Emily's house. What is the distance between James's house and Joan's house?

4. What is the expression that describes the number that is located at a distance of 20 from the right of (-7)?

Ⓐ $(-7) + 20$ Ⓑ $20 + 7$

Ⓒ $(-7) - 20$ Ⓓ $20 - (-7)$

Ⓔ $7 - 20$

1. The highest recorded temperature at one city is $120^{\circ}F$ and the lowest temperature recorded at the same city is $-28.3^{\circ}F$. What is the difference of those two temperatures?

2. The freezing point of one liquid is $-50^{\circ}C$ and the freezing point of another liquid is $-117^{\circ}C$. What is the difference between the two freezing points?

3. At noon, the temperature was $62^{\circ}F$. Twelve hours later, the temperature dropped by $123^{\circ}F$. What was the temperature twelve hours later?

4. Dorothy found that Philip's house is 160.5 meters west from her house and Paul's house is 57.7 meters east from her house. What is the distance between Philip's house and Paul's house in meters?

1. What is $9\frac{2}{11} + 10\frac{10}{11}$ in a fractional form?

2. Find $35\frac{11}{14} - 21\frac{13}{14}$.

3. At James's birthday, he received a $40 gift certificate. He used the gift certificate to buy 3 used video games whose prices were $5.69, $7.79, and $6.99. What is the remaining balance on the gift certificate?

4. Evaluate $7.74 - (-3.84) + \frac{53}{100}$.

1. Evaluate 290×0.5.

2. Divide 38.4013 by 11.9.

3. Rachel needs $4\frac{2}{7}$ cups of flour to bake a cake. If she only has a $\frac{1}{7}$-cup measuring cup, how many times will she need to fill the measuring cup?

4. The temperature generally drops at a rate of about 2.5 degrees per 2000 feet increase in altitude. If the temperature at the ground level is $25^{o}F$, what is the temperature at 10000 feet above the ground level?

1. What is $78 \times \left(-\dfrac{1}{12}\right)$? Write the answer in decimal form if needed.

2. What is $(-1) \times 3.7 \times (-6) \times (-3)$?

3. What is the area, in square miles, of a rectangular park with 2.8 miles long and 1.6 miles wide?

4. James is planning to put tiles on the flood in his bathroom which has a length of 6 feet and a width of 7 feet. If one square foot tile costs $3.11, what is the total cost in dollars?

1. What is the quotient of $(-80) \div (-16)$?

2. Which of the following would have a different quotient with others?

Ⓐ $\dfrac{-4}{5}$ Ⓑ $\dfrac{-12}{-15}$ Ⓒ $-\dfrac{16}{20}$

Ⓓ $\dfrac{-8}{10}$ Ⓔ $\dfrac{4}{-5}$

3. If the weight of 17 boxes of papers is 263.5 pounds, what is the weight of each box in pounds?

4. There are 106 water bottles on a table. A student wants to put all the bottles into boxes which can hold 17 water bottles each. How many box(es) does he or she need?

1. Evaluate $-\dfrac{5}{6} \times \left(-\dfrac{2}{7}\right)$.

2. Evaluate $\dfrac{4}{5} \times \left(\dfrac{3}{5} \div (-5)\right)$.

3. Which of the following is same as 2.1×4.35?

Ⓐ $21 \times 435 \times 1000$ Ⓑ $21 \times 435 \times 10$

Ⓒ $21 \times 435 \div 1000$ Ⓓ $21 \times 435 \div 10$

Ⓔ $21 \times 435 \div 100$

4. What is $\dfrac{1}{1 + \dfrac{1}{3}}$ in fractional form?

---

1. Which of the following is _not_ a rational number?

Ⓐ $\frac{3}{2}$

Ⓑ 9.01212121.... repeating

Ⓒ −2.13424533.... non repeating

Ⓓ −2.8

Ⓔ 0

2. Find the quotient of 1.781 ÷ 1.3.

---

3. What is a decimal form of a number, $\frac{23}{3}$? Round the answer to the nearest hundredth.

4. Ian organized his drawer and found that $\frac{8}{25}$ of his clothes are jeans. Write this fraction in decimal form.

1. The cost of **32** movie tickets is **$212.16**. How much does each ticket cost?

2. Megan bought a pack of **7** pairs of socks. If the pack costs **$27.79**, what is the cost for a pair of socks?

3. A bag contains **99** cups of cat food. Rachel's cat eats $1\frac{3}{8}$ cups a day. How many days can she feed her cat?

4. Megan bought $36\frac{1}{5}$ feet of ribbon to decorate her room which is square shaped. If she put ribbon straight around the room with $5\frac{2}{7}$ feet one side, how many feet of ribbon would be left? Write the answer using an improper fraction.

1. In the desert, the temperature at noon is $86°F$. It dropped $86°F$ at midnight. What is the temperature at midnight in F?

Ⓐ $-43°F$ Ⓑ $0°F$ Ⓒ $-172°F$

Ⓓ $86°F$ Ⓔ $-86°F$

2. At one spot, the day time temperature had reached to $18°F$, and the night time temperature had dropped to $(-21)°F$. what expression describes the temperature change in F?

Ⓐ $(-21) - 18$ Ⓑ $18 - 21$

Ⓒ $21 - 18$ Ⓓ $18 - (-21)$

Ⓔ $18 + (-21)$

3. Two cars are leaving the same parking lot in opposite directions. One car is heading west with a speed of 60 mph, and the other car is heading east with a speed of 65 mph. What is the distance between two cars after 3 hours?

Ⓐ 375 miles   Ⓑ 128 miles

Ⓒ 245 miles   Ⓓ 255 miles

Ⓔ 125 miles

4. If the highest point on a continent is 3.3 kilometers above sea level and the lowest point is 0.15 kilometers below sea level, what is the difference between these elevations in kilometers?

Ⓐ 1.8   Ⓔ 3.45   Ⓒ 3.15

Ⓓ 3.6   Ⓔ 4.8

5. Emily's class is doing an art project and she has 41.61 yard of string. If she wants to give each of her 19 students an equal amount of string, how much will each student get?

Ⓐ 1.095   Ⓑ 4.38   Ⓒ 2.19

Ⓓ 1.79   Ⓔ 2.59

6. Ethan's school is going on a field trip and school buses will be rented for 142 students, 10 chaperons, and 4 teachers. If the seating capacity of a school bus is 49, how many school buses must be rented?

Ⓐ 2   Ⓑ 6   Ⓒ 5

Ⓓ 4   Ⓔ 3

7. Which of the following has the same quotient as $8.35 \div 4.5$?

Ⓐ $835 \div 45 \div 1000$ Ⓑ $835 \div 4.5$

Ⓒ $835 \div 45 \times 1000$ Ⓓ $835 \div 45$

Ⓔ $83.5 \div 45$

8. Philip's weight is $1\frac{1}{9}$ times Andrew's weight, and Andrew's weight is $\frac{10}{11}$ times Jeff's weight. If Philip's weight is $33\frac{1}{3}$ kg, what is Jeff's weight in kg?

Ⓐ $33\frac{8}{9}$ kg   Ⓑ $33\frac{1}{9}$ kg   Ⓒ 34 kg

Ⓓ 32 kg   Ⓔ 33 kg

9. What is $(-16) - (-7)$

Ⓐ -9 Ⓑ -23 Ⓒ 9

Ⓓ 15 Ⓔ 23

10. Jeff drove 234.9 miles for 4 hours 30 minutes, and his car consumes 0.135 gallon for one mile. If he keeps the same speed and used 28.188 gallons, how many hours did he drive?

Ⓐ 4   Ⓑ 2   Ⓒ 5

Ⓓ 3   Ⓔ 6

11. The checking account balance was initially $123. Then a man used a debit card associated with the checking account 5 times, and he spent $127 each time. How much money is left in his checking account?

Ⓐ $-$512$ Ⓑ $-$758$ Ⓒ 758

Ⓓ 512 Ⓔ 250

12. A car travels 18.718 miles on 1.4 gallon of gas. If 1 gallon of gas costs $1.77, how much does it cost to travel 93.59 miles?

Ⓐ 17.70 Ⓑ 12.39 Ⓒ 35.40

Ⓓ 123.90 Ⓔ 24.78

13. What is the value of $-10 - \{(10 - 5) - 14\}$?

Ⓐ -1 Ⓑ 19 Ⓒ -11

Ⓓ -9 Ⓔ -39

14. Simplify the following expression.

$$\left(\frac{8}{7}\right) \div \left(-\frac{8}{7}\right) \times \left(-\frac{3}{10}\right) \div \left(+\frac{7}{9}\right)$$

Ⓐ $-\dfrac{27}{70}$ Ⓑ $\dfrac{21}{10}$ Ⓒ $-\dfrac{21}{10}$

Ⓓ $\dfrac{27}{70}$ Ⓔ $\dfrac{10}{9}$

15. What is the product of two numbers when both of them are 10 away from -1 on the number line?

Ⓐ -108 Ⓑ -120 Ⓒ -99

Ⓓ -110 Ⓔ 0

16. What is the value of x if $x \div (-4.9) = 34$?

Ⓐ 1666 Ⓑ -16.66 Ⓒ 166.6

Ⓓ -1666 Ⓔ -166.6

Expressions and Equations

7.EE

1. Simplify the expression, $8x + 5 - 11x$.

2. Factor the expression, $-35x - 63y$.

3. If $a = 6x + 2$ and $b = 8x - 9$, what is $a - b$?

4. What is the product of the coefficients of a^2 and a when we simplify the expression $(4a^2 - 3a + 9) - 6(a^2 - 6a - 4)$?

1. A rectangular mat is 7 times as long as its width. What is the simplified expression for the perimeter of the mat in terms of the width w?

2. The bill for Emily's dinner was x dollars. If she added a 16% tip, what is the expression that shows the total amount she paid?

3. An item originally priced at b dollars is marked 24% off. What is the expression of the sales price?

   Ⓐ $b - 24$ Ⓑ $b - 0.24b$

   Ⓒ $b - 0.76b$ Ⓓ $b - 0.24$

   Ⓔ $b - 24b$

4. At a restaurant, Matthew ordered lunch that costs f dollars. If he left a 16% tip and the sales tax 6.5%, which of the following shows the total cost?

   Ⓐ $f + 0.16 + 0.065f$ Ⓑ $f + 0.16f + 0.065$

   Ⓒ $f + 0.16f + 0.65$ Ⓓ $f + (0.16 + 0.065)f$

   Ⓔ $f + (0.16 \times 0.065)f$

1. If 14 more than twice a number is −22, what is the number?

2. An artist wants to draw on a canvas where the length of the canvas is 8 less than twice the width. If the perimeter of the canvas is 92 inches, what is the length of the canvas in inches?

3. The size of a rectangular garden is 50 feet by 55 feet. If there is a sidewalk around the garden that is 2 feet wide, what is the outside perimeter of the sidewalk, in feet?

4. Bernie lowered the temperature of a room by $\frac{1}{4}$. Then Gus raised the temperature by 11%. What percent of the beginning temperature of the room is the new temperature?

1. A book store sells paperbacks for $8 each, and you will receive a $3 discount if you spend at least $25. What is an inequality that represents the least number of paperbacks you have to buy to get the discount when n represents the number of books?

2. Emily wants to send an item that is 34 pounds using an air mail. What is the inequality that represents the weight of a bag she can buy if the total weight cannot be greater than 50 pounds? Assume w represents the weight of a bag in pounds.

3. Brian will be 6 years less than twice of Ethan's age in 8 years. Which of the following is the expression of Brian's age in 8 years if Ethan's current age is n?

   Ⓐ $2n + 20$ Ⓑ $2n + 10$

   Ⓒ $2n + 4$ Ⓓ $2n + 2$

   Ⓔ $2n + 14$

4. A total 297 students went on a field trip. If 5 buses were filled and 7 students traveled in cars, how many students were in each bus?

1. At a large farm, a bag of onions weighs 25 pounds. If a wheelbarrow can carry up to 400 pounds, what is the greatest number of bags the wheelbarrow can carry?

2. Solve the equation, $0 = \dfrac{t}{6} + 12$.

3. After spending $\dfrac{5}{8}$ of his monthly allowance, he has $15 left. What was his monthly allowance in dollars?

4. To fix a car, a mechanic charges $28 for each hour of work, and the new parts cost $168. The total cost of fixing the car is $392. How many hours did it take the mechanic to fix the car?

1. Solve the inquality, $x - 19 \leq -24$.

2. Solve the inequality, $-2x + 10 < -6$.

3. Julie solved the inequality and found the following as a graph of solutions. Which of the following inequalities did she solve?

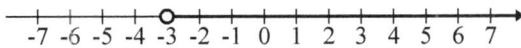

-7 -6 -5 -4 -3 -2 -1 0 1 2 3 4 5 6 7

Ⓐ $2x + 19 > 25$ Ⓑ $2x + 25 \geq 19$

Ⓒ $2x + 19 < 25$ Ⓓ $2x + 25 > 19$

Ⓔ $2x + 25 < 19$

4. How many integers for x satisfy this inequality, $|4x - 8| \leq 4$?

1. The area of a triangle is $16cm^2$. What is the height of a triangle when the base is 8cm?

Ⓐ 3 cm   Ⓑ 2 cm   Ⓒ 6 cm

Ⓓ 7 cm   Ⓔ 4 cm

2. Kevin had 20 apples and Ethan had 1.5 times more apples than Kevin. If Kevin gives Ethan 4 apples, what is the ratio of Ethan's apples to Kevin's apples rounded to the nearest hundredth?

Ⓐ 2.5   Ⓑ 4.6   Ⓒ 3.13

Ⓓ 3.38   Ⓔ 2.7

3. The number of ants in an anthill increases by 22 percent in one week. In the next week, the population of the anthill decreases by $\frac{1}{5}$. What percent of the original number of ants is the current population of the anthill?

Ⓐ 97.6%   Ⓑ 106.4%   Ⓒ 88.8%

Ⓓ 80%   Ⓔ 71.2%

4. If the sum of three consecutive numbers is 246, what is the largest of these numbers?

Ⓐ 85   Ⓑ 83   Ⓒ 82

Ⓓ 86   Ⓔ 81

5. Find the number of integer values for x that satisfy the inequality, $|2x - 8| < 4$.

Ⓐ 6   Ⓑ 2   Ⓒ 3

Ⓓ 5   Ⓔ 4

6. At an online store, the price of a shirt was increased by 15% and the shipping and handling cost was $5. The total amount Brian paid was $31. What was the original price of a shirt before it was increased? Round appropriately if necessary.

Ⓐ $18.26   Ⓑ $26   Ⓒ $22.61

Ⓓ $23.61   Ⓔ $26.61

7. Jessica is selling 6 handmade bracelets, and the cost to make one bracelet is $5.67. What is the minimum price of one bracelet if she wants to have a profit more than $27 when she sells all of bracelets?

Ⓐ $21.33   Ⓑ $32.67   Ⓒ $11.17

Ⓓ $5.45   Ⓔ $10.18

8. Matthew has $19 to spend at the mall. If he plans to spend $2 on snacks, and plans to buy key chain which costs $2 each, what is the maximum number of key chains he can buy?

Ⓐ 11   Ⓔ 10   Ⓒ 9

Ⓓ 7   Ⓔ 8

9. There are 4 cups, A, B, C, D. After pouring water in cup A, you pour 25 oz more water in cup B than A, and 25 oz more water in cup C than B, and so on. If the total amount of water in 4 cups is 670 oz, how many oz of water is in cup D?

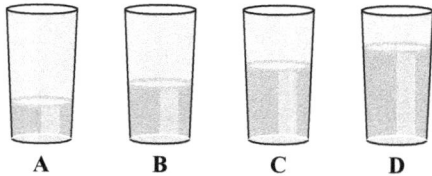

A B C D

Ⓐ 205   Ⓑ 155   Ⓒ 180

Ⓓ 130   Ⓔ 230

10. A store reduced the price of shirts by 11 percent. Then, they raised price by $\frac{1}{4}$. What percent of the original price is the current price?

Ⓐ 111.25%   Ⓑ 118.125%   Ⓒ 131.875%

Ⓓ 104.375%   Ⓔ 125%

11. What is the simplest form of the expression,
$$\frac{4x + 2y}{8x + 4(2x + 3y) - 4y} \times 12 + 2x \ ?$$

Ⓐ $12 + 2x$ Ⓑ $3 + 2x$ Ⓒ $6x + 2y$

Ⓓ $4 + 2x$ Ⓔ $5x + 2y$

12. A factory produces cups and each cup weighs 87.6g. The manager put 34 cups in a box and measured the weight of the box. It was 3204.4g. Then, he took some cups out and measured the weight of the box again, and it was 1627.6g. How many cups did he take out from the box?

Ⓐ 29   Ⓑ 18   Ⓒ 33

Ⓓ 19   Ⓔ 10

Geometry

7.G

1. Ian is drawing the gym which is 49 feet wide. If it is 7 inches wide in the drawing, how many feet is represented by 1 inch?

2. Kristy is planning to visit her grandmother. She found that the distance between her and her grandmother's house is 7.5 inches on the map. If a map uses a scale of 1 inch = 14 miles, how far is her grandmother's house, in miles?

3. Megan has a(n) 20 : 1 scale of the floor plan of her house. If the dimensions of her rectangular bed room are $1\frac{1}{2}$ inches by $2\frac{3}{10}$ inches on the floor plan, What is the area of her real bed room in square feet? Round the answer to the nearest square feet if necessary.

4. On a blueprint, 2 inches correspond to 3 feet. The size of living room on the blueprint is 8 inches by 5 inches. What is the actual size of the living room?

1. Two sides of a triangle are given as **11** and **14**. Which of the following can *not* determine a triangle?

Ⓐ 7   Ⓑ 11   Ⓒ 24

Ⓓ 15   Ⓔ 2

2. If two interior angles of a triangle are **37°** and **33°**, what is the third interior angle?

3. Which of the following angles can form a quadrilateral when three other angles are given as 129°, 54°, and 96° ?

Ⓐ 129°   Ⓑ 81°   Ⓒ 96°

Ⓓ 90°   Ⓔ 54°

4. Which of the following conditions can *not* be a triangle?

Ⓐ Angles of 72°, 50°, and 58°

Ⓑ Side lengths of 5, 7, and 3

Ⓒ Angles of 90°, 50°, and 40°

Ⓓ Side lengths of 7, 5, and 18

Ⓔ Side lengths of 8, 6, and 13

1. What is the shape of a cross section when a cylinder is cut perpendicular to its base?

2. Which of the following solids has no vertex?

   Ⓐ Rectangular Pyramid   Ⓑ Rectangular Prism

   Ⓒ Triangular Pyramid   Ⓓ Sphere

   Ⓔ Cube

3. When a solid is cut parallel to the base, the cross section is a triangle. When a solid is cut perpendicular to the base, the cross section is a rectangle. What is this solid?

4. Which of the following objects always has the same two dimensional figure when the object is sliced at any direction?

   Ⓐ Sphere   Ⓑ Pyramid

   Ⓒ Rectangular prism   Ⓓ Cube

   Ⓔ Cylinder

1. Christine bought a package of plates for the birthday party. If the diameter of a plate is 18 cm, what is the area of a plate in square cm? Use 3.14 for π.

2. What is the area of a circle when the diameter of a circle is 16?

3. Each circular table at a restaurant has a plastic strip around the edge. If the radius of a table is 1.5 feet, what is the length of a plastic strip in feet? Use 3.14 for π.

4. What is the area of a circle with a circumference of 6π ?

1. If two angles, 65° and $(5x + 5)^o$, are complementary, what is the value of x?

2. What is the value of $y - x$ in the figure below?

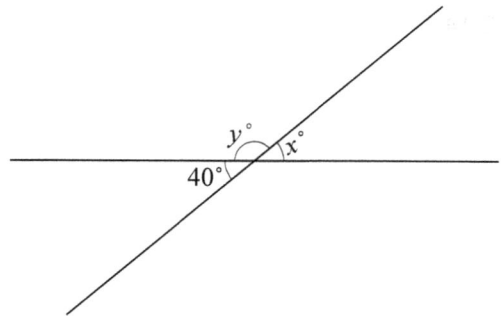

3. If $\angle a = 69$, what is $\angle c$, in degree?

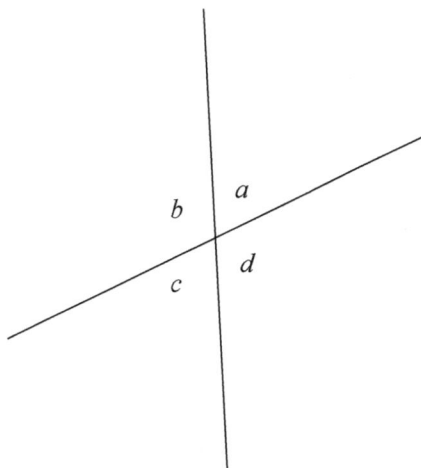

4. Triangles ABC is isosceles triangle with the side lengths of AB and AC which are equal. What is the measurement of the angle BAC?

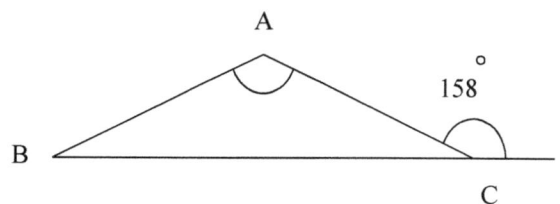

1. What is the surface area of a rectangular box, in square centimeters, if the base is a square with a side length 7 cm and the height 4 cm?

2. What is the volume, in cubic centimeters, of a rectangular prism whose dimensions are 8 cm by 6 cm by 7 cm?

3. Gregory is pouring water in the cylinder can that has a radius of 7 inches and a height of 5 inches. What is the maximum amount of water, in cubic inches, that can fill the can?
Use 3.14 for π.

4. Gregory is making a large wooden toy box that is 2 feet wide, 2.5 feet long, and 1.5 feet height. If the wood costs $2.61 per square foot, what is the total cost of the wood to make the toy box? Round the answer to the nearest cent.

1. A model of a tall tower uses a scale 1 : 16. If the height of the actual tower is 84.8 feet, what is the height, in feet, of the model?

   Ⓐ 6.1   Ⓑ 5.4   Ⓒ 6.7
   Ⓓ 5.2   Ⓔ 5.3

2. If the circumference of a circle is 6π, what is the area of the circle?

   Ⓐ 36π Ⓑ 9   Ⓒ 18π
   Ⓓ π Ⓔ 9π

3. Find the circumference of a circle when the radius of a circle is 22.

   Ⓐ 44π Ⓑ 22π Ⓒ 88π
   Ⓓ 242π Ⓔ 484π

4. The volume of a rectangular swimming pool is 882 cubic feet with a height 7 feet. If the length is twice the height, what is the width of the swimming pool in feet?

   Ⓐ 4.5   Ⓑ 9   Ⓒ 18
   Ⓓ 2.25   Ⓔ 36

5. Two angles, 51° and $(35 + 2x)^o$, are supplementary. What is the value of x?

Ⓐ 2 Ⓑ 5 Ⓒ 47

Ⓓ 42 Ⓔ 48

6. What is the value of x in the figure below?

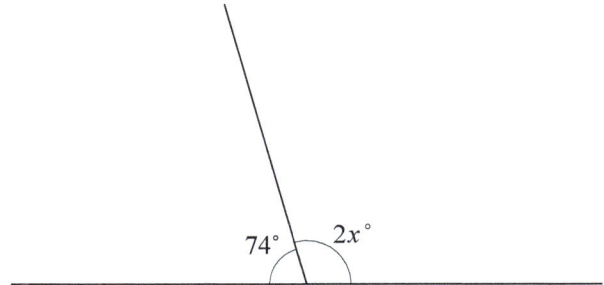

$74°$ $2x°$

Ⓐ 81 Ⓑ 91 Ⓒ 53

Ⓓ 106 Ⓔ 86

7. Christine is wrapping a gift box that is 10 inches by 6 inches by 4 inches. What is the least amount of wrapping paper she needs in order to wrap the gift box?

Ⓐ 240 cubic inches Ⓑ 186 square inches

Ⓒ 120 cubic inches Ⓓ 124 square inches

Ⓔ 248 square inches

8. What is the value of x?

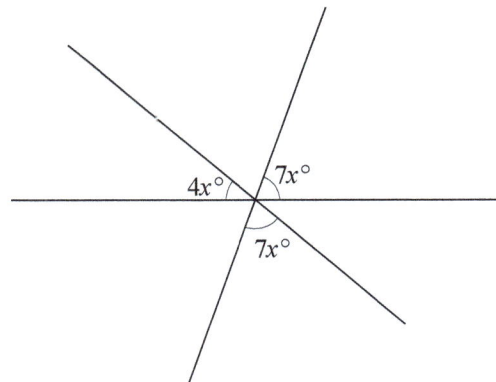

$4x°$ $7x°$

$7x°$

Ⓐ 13 Ⓑ 8 Ⓒ 28

Ⓓ 5 Ⓔ 10

Statistics and Probability

7.SP

Skill Practice: Understand that statistics can be used to gain information about a population using sample (7.SP.A.1)

Name: _____

Date: _____

1. Statistics can be used to gain information about a _____ by examining a _____ of the population. What are two words (in order) that should go in the blanks?

   Ⓐ result, data

   Ⓑ population, sample

   Ⓒ sample, data

   Ⓓ population, data

   Ⓔ data, result

2. Which of the following sampling methods best represents the opinion of students in one class?

   Ⓐ Choose 14 students who volunteer

   Ⓑ Choose 14 students who sit in the front

   Ⓒ Choose 14 students randomly

   Ⓓ Choose 14 male students

   Ⓔ Choose 14 students who wear glasses

3. Which of the following is an unbiased sample for colors of leaves in October?

   Ⓐ 60 leaves on maple trees and 60 leaves on oak trees

   Ⓑ 120 leaves on trees

   Ⓒ 120 leaves on maple trees

   Ⓓ 120 leaves on the ground

   Ⓔ 60 leaves on trees and 60 leaves on the ground

4. Emily is doing a survey for the most popular ice cream flavor in her school. Which of the following samples is likely to be representative of all the students in her school?

   Ⓐ 55 random students

   Ⓑ 55 students in her grade

   Ⓒ 55 students in the library

   Ⓓ 55 students in the gym

   Ⓔ 55 students in her class

1. Below is the number of words of five randomly selected pages of a 170 page book. Estimate the total number of words of this book.

| 10, | 12, | 15, | 12, | 11 |

2. Below is the number of words of each five randomly selected pages of two different books. Which of the following is mostly true?

Book 1: 33, 34, 36, 37, 39
Book 2: 17, 21, 25, 30, 36

(A) The variation in estimates of the number of words in Book 2 is greater than those in Book 1.

(B) The estimates for the number of words in Book 2 is greater than those in Book 1.

(C) The estimated total number of words in Book 1 is 895.

(D) The estimates of the number of words in Book 1 varies more than those in Book 2.

(E) The estimated total number of words in Book 2 is 129.

3. John wants to predict the winner of student president election based on survey data. Which of the following provides the best prediction (i.e., the most generalizable prediction)?

(A) Conduct survey randomly selected 10 students each in the biggest 5 classes

(B) Conduct survey the first 10 students from each of 5 geography classes

(C) Conduct survey randomly selected 10 students in 5 classes John is taking

(D) Conduct survey the first 5 students from each of 10 geography classes

(E) Conduct survey randomly selected 5 students from each of 10 classes randomly selected

4. Ethan wants to make an inference about the mean (average) number of paragraphs in each page of a geometry textbook. Which of the following is the most relevant method to accurately draw such inference?

(A) Compute the mean of paragraphs from 10 ramdomly selected chapters

(B) Compute the mean of paragraphs from 20 ramdomly selected pages

(C) Compute the mean of paragraphs from the first 20 pages

(D) Compute the mean of paragraphs from 10 ramdomly selected pages

(E) Compute the mean of paragraphs from the last 10 pages

1. James surveyed heights of players in the gymnastics team and squash team. James found that the mean of the height of players in the gymnastics team is 1.8 meters with standard deviation .60, and the mean of heights of players in the squash team is 1.90 meters with standard deviation .60. Which of the following is the most appropriate interpretation on this survey result?

Ⓐ The mean height of players on the squash team is greater than gymnastics team as much as 0.6 times the standard deviation

Ⓑ The mean height of players on the gymnastics team is 0.1 meter greater than the squash team

Ⓒ The mean height of players on the gymnastics team is 1.8 meter greater than the squash team

Ⓓ The mean height of players on the squash team is 0.7 meter greater than the gymnastics team

Ⓔ The mean height of players on the squash team is greater than the gymnastics team as much as $\frac{1}{6}$ times the standard deviation

2. The following chart shows the heights of volleyball players and ice hockey players. What inference can be made based on this chart?

Volleyball	Ice Hockey
5'9"	5'5"
5'10"	5'5"
6'3"	5'8"
6'3"	5'9"
6'6"	6'2"
6'7"	6'2"

Ⓐ No inference can be made.

Ⓑ Volleyball players have a greater average appetite

Ⓒ Volleyball players have a higher average height

Ⓓ Ice hockey players have a lower average weight

Ⓔ Ice hockey players have a greater average running speed

3. The following plots show the number of minutes each student spent reading and playing games in a day. Which statement is incorrect?

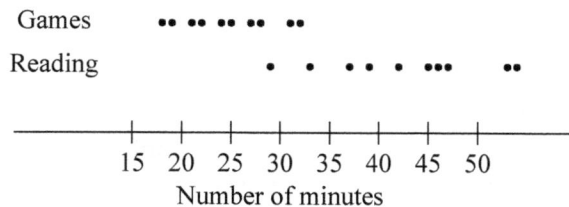

Games   •• •• •• •• ••

Reading   • • • • • ••• ••

15 20 25 30 35 40 45 50
Number of minutes

Ⓐ The average minute spent reading is greater than the average minute spent playing games

Ⓑ Minutes spent reading are more spread out

Ⓒ Minutes spent reading varies more than minutes spent playing games

Ⓓ On average, students spend more minutes reading than playing games

Ⓔ The range of minutes spent reading is less than the range of minutes spent playing games

4. The graphs below show distributions of Math exam and English exam scores. Which statement is correct?

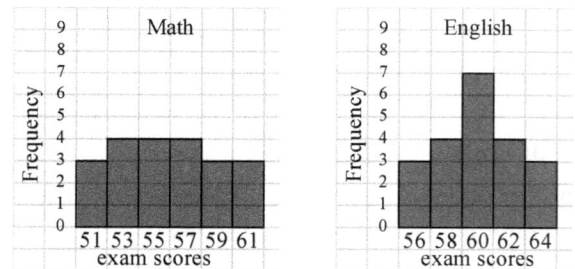

Math — Frequency (0–9), exam scores 51 53 55 57 59 61

English — Frequency (0–9), exam scores 56 58 60 62 64

Ⓐ Math exam scores vary more than English exam scores

Ⓑ Math and English exam scores have the same range

Ⓒ The highest score is in the Math exam

Ⓓ The average of Math exam scores is higher than English exam scores

Ⓔ Math exam scores have a higher mode

1. The algebra book has total 120 pages and the music book has total 110 pages. James randomly selected 40 pages from both books and found the mean of the number of words in each page of the algebra book is 500 with a standard deviation 60, while it is 390 words with a standard deviation 60 in the music book. Based on these, which of the following is the most appropriate?

   Ⓐ The number of words in the algebra book is less than music book

   Ⓑ The number of words in each page of the algebra book is generally more than music book

   Ⓒ There are 60000 words in the algebra book

   Ⓓ The estimated number of words in the music book is 60

   Ⓔ The estimated number of words in the algebra book is 500

2. The box plots for the data are given below. Which of the following is a correct statement?

   Ⓐ Data set X spreads more than data set Y

   Ⓑ The first quartile for data set X is higher than the first quartile for data set Y

   Ⓒ The maximum value of data set X is lower than the maximum value of data set Y

   Ⓓ Data set X has a lower mean than data set Y

   Ⓔ Data set X has higher median than data set Y

3. The following plots show the number of minutes each student spent reading and playing games in a day.

   Which pieces of information can be gathered from these dot plots? Select all that apply.

   a) The minutes spent reading were more on average than playing games.
   b) Every students spent more time reading than playing games.
   c) The minutes spent reading vary more than playing games.

4. The following graphs show the test scores for Chemistry and Biology. Which statement is incorrect?

   Ⓐ Chemistry test scores and biology test scores vary the same

   Ⓑ On average, chemistry test scores are higher than biology test scores

   Ⓒ The highest score is from chemistry test

   Ⓓ Chemistry test scores are always lower than biology test scores

   Ⓔ The range of chemistry test scores is 24

1. Which of the following is the _least_ appropriate statement?

   Ⓐ The probability of 0 represents that an event never happens

   Ⓑ The probability of an event is a number between −1 and 1

   Ⓒ The probability 0.6 represents more likely event than that of 0.4

   Ⓓ The probability of 1 represents that an event always happens

   Ⓔ The probability 0.5 represents the chance of an event occurring is the same as it is not occurring

2. Which of the following numbers could not represent probability?

   Ⓐ 1   Ⓑ 0.8   Ⓒ −0.8

   Ⓓ $\frac{4}{5}$ Ⓔ 0

3. A bag contains red, blue, purple, yellow, and green marbles. The probability of randomly choosing a blue marble is 5%, a purple marble is 15%, a yellow marble is 27%, and a green marble is 23%. Which color marble is most likely to be chosen?

4. A bag contains red and blue marbles, and the probability of randomly choosing a red marble is 40%. How many blue marbles are in the bag if there is a total of 95 marbles in the bag?

1. If Paul is rolling a number cube (fair die with 1 to 6 as outcomes) 900 times, which of the following is the _least_ appropriate statement on this experiment?

   Ⓐ Paul can predict that the chance of 1 or 4 rolled is $\frac{1}{3}$

   Ⓑ Paul can predict that 4 would be rolled roughly 150 times

   Ⓒ Paul can predict that 1 or 4 would be rolled roughly 300 times

   Ⓓ Paul can predict that 1 or 4 would be rolled exactly 300 times

   Ⓔ Paul can predict that 1 would be rolled roughly 150 times

2. If you flip two coins 20 times, how many times will both coins likely land on heads?

3. If Ethan is rolling a fair die, how many times can he predict to roll 4 or a number that is greater than 4 after 66 rolls?

4. In the school election, 49% students supported candidate "A" and 23% students supported candidate "B". What is the probability that a randomly selected student would have supported neither candidate "A" nor candidate "B"?

1. The chart below shows how much Jim spent in a month. He earns $2600 a month after tax. How much more does he spend on groceries than savings?

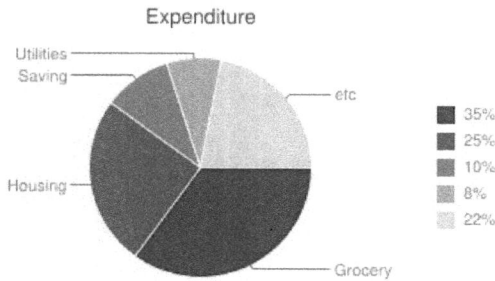

Expenditure

Utilities
Saving
etc

Housing

Grocery

- 35%
- 25%
- 10%
- 8%
- 22%

2. Teacher Paul wants to randomly select a student his class with equal chance of being selected. If the class has 40 students, which of the following is the *least* appropriate statement?

(A) The shape of the distribution will be rectangular

(B) The chance of a student being selected is $\frac{1}{40}$ and this is same for all students in this class

(C) Sum of the chances for each student being selected will decrease as the class size increase

(D) This is a uniform probability model

(E) The chance of a student being selected will decrease as the class size increase

3. There are 19 balls in a bag. 4 of them are red balls, 2 are blue balls, and the rest are yellow balls. What is the probability of drawing a yellow ball if one ball is drawn randomly?

4. The darts target is given below. When one dart is thrown, what is the probability that one dart hits a factor of 12?

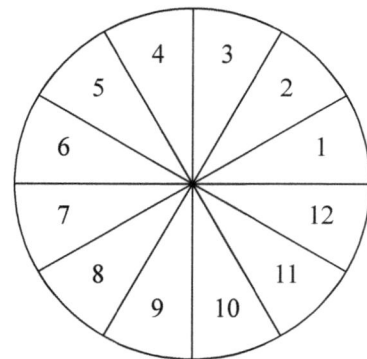

1. Merchant Gregory wants to randomly select a customer each of whom has an equal chance of being selected in his store. If there are 40 customers in the store, which of the following is the most appropriate statement?

Ⓐ The chance of a customer being selected will increase as the number of customer increase

Ⓑ Sum of the chances for each customer being selected will decrease as the store size increase

Ⓒ The chance of a customer being selected is $\frac{1}{40}$ and this is same for all customers in this store

Ⓓ This will be a bell shaped curve probability model

Ⓔ If Gregory attempts this selection 40 times, customer Rachel in the store will be selected at least once

2. Gregory found that he has 6 white t-shirts out of 22 t-shirts. There is an equal chance of selecting one of the 22 t-shirts from the dresser. What is the probability that the t-shirt he randomly picks is white?

3. There are math, history and physics books on a shelf. If the probability of picking a math book is $\frac{1}{9}$ and the probability of picking a physics book is $\frac{1}{8}$, what is the probability of picking a history book?

4. Dorothy has 160 books in her room. There are 90 fiction books and 70 nonfiction books. What is the probability that she randomly pick a nonfiction book?

1. Brian found a coin in his backyard, and he wanted to find out the probability of heads and/or tails appearing when spinning the coin. After spinning the coin 90 times, there were 40 heads and 50 tails. Based on this information, which of the following is the _least_ appropriate statement?

   Ⓐ The chance of tails appearing would be approximately $\frac{50}{90}$

   Ⓑ The accuracy of estimation for the chance of heads will increase as the number of spinnings increase

   Ⓒ The chance of tails appearing is $\frac{50}{90}$

   Ⓓ The chance of heads appearing is estimated as $\frac{40}{90}$

   Ⓔ The chance of heads or tails is possible to be same

2. Thomas found a coin in his backyard, and he wanted to find out the probability of heads or tails. After spinning the coin 95 times, he noticed 35 times of heads and 60 times of tails. Based on this information, which of the following is the most appropriate statement?

   Ⓐ The distribution of heads or tails probability is not a rectangular shape

   Ⓑ The chance of heads appearing is $\frac{35}{95}$

   Ⓒ The chance of tails appearing would be approximately $\frac{60}{95}$

   Ⓓ The chance of heads will increase as the number of spinnings increase

   Ⓔ Sum of the chances for heads or tails will decrease as the number of spinnings increase

3. In a color pencil box, there are 6 blue color pencils, 4 red color pencils, and 9 green color pencils. If one color pencil is chosen randomly, what is the probability of choosing either blue or green color pencil?

4. There are 5 red balls, 4 blue balls, and 6 yellow balls. If one ball is chosen randomly, what is the probability that the ball is not yellow?

1. A fair six-sided die is rolled twice. What is the theoretical probability that the first number that comes up is less than or equal to the second number?

2. When rolling two dice, what is the probability that the sum of two numbers is less than or equal to 9?

3. If you flip three fair coins, what is the probability that you'll get tails on the first two flips and heads on the last flip?

4. A restaurant serves 6 kinds of pasta (spaghetti, bow ties, fettuccine, penne, rotini, and macaroni) in 4 different flavors (tomato sauce, cheese sauce, olive oil, and Alfredo sauce). If you randomly pick one pasta and flavor, what is the probability that you'll order a pasta that is not tomato macaroni?

1. Rachel rolls a die and flips a coin. What is the probability of rolling a **6** and the coin landing on a tail? Answer in fraction form.

2. James spins two coins. Both coins have a 60% chance of heads appearing. If James spins two coins 400 times, which of the following is the most appropriate statement?

   Ⓐ Each spinning of two coins has 0.6% chance of having both coins show heads

   Ⓑ It is impossible that all spinnings show both heads

   Ⓒ If the first coin occurs heads, the chance of getting heads in the second coin would be 40%

   Ⓓ The number of spinnings that both show heads is **96**

   Ⓔ There is **36%** chance of two heads appearing on both coins

3. When rolling two dice, what is the probability that the difference between two numbers is 3?

4. A fair six-sided die is rolled twice. What is the theoretical probability that the first number rolled is the same as the second number rolled?

1. What is the sample space for tossing a coin and a fair four-sided dice?

   Ⓐ {HH, TT, 1, 2, 3, 4}

   Ⓑ {H1, H2, H3, H4, T1, T2, T3, T4}

   Ⓒ {H, T, 1, 2, 3, 4}

   Ⓓ {H1, H2, H3, H4, H5, H6, T1, T2, T3, T4, T5, T6}

   Ⓔ {HH, HT, TH, TT, 1, 2, 3, 4}

2. How many outcomes are possible when two fair six-sided dice are rolled?

3. If you roll a six-sided die and toss a coin, how many outcomes are possible?

4. You have three pairs of pants, four shirts, and four pairs of socks. How many different ways can you dress up if you choose one pair of pants, one shirt, and one pair of socks?

1. The figures below present the donors blood types and genders in Doctor Brian's clinic. Assuming the blood type distribution is equal for female and male, and there are currently 500 donors in the clinic, which of the following is the best approximation of the number of male donors with blood type B?

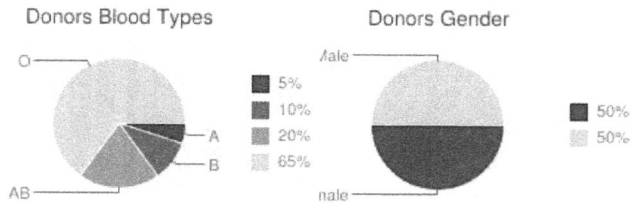

Donors Blood Types

Donors Gender

Male

- 5%
- 10%
- 20%
- 65%

O

A

B

AB

male

- 50%
- 50%

Ⓐ 27

Ⓑ 25

Ⓒ 28

Ⓓ 23

Ⓔ 29

2. There are 200 students who are playing sports at one high school. The following charts show the percentage of students who are playing each sport and the percentage of male and female students in the high school.

Sports	Percent	Male	Female
Baseball	20%	60%	40%
Basketball	20%		
Soccer	15%		
Volleyball	45%		

What is the approximated number of female students who play baseball?

3. Teacher James purchased lottery tickets from a local store for a school event. Each lottery ticket has 20% chance of winning, and all 500 event participants get two tickets. Based on this information, which of the following is the most appropriate statement?

Ⓐ Each particpants has 0.2% chance of winning both lotteries

Ⓑ It is impossible that all participants can win both lotteries

Ⓒ Each particpants has 4% chance of winning both lotteries

Ⓓ If a participant wins the first lottery, the chance of winning the second lottery would be 80%

Ⓔ The number of participants who wins both lotteries is 80

4. There are 3 defective products out of 14 in a box. If two products were selected without replacement, what is the probability that both products are defective?

1. Which of the following is an unbiased sample of students in Corner middle school?

   Ⓐ 60 students in PreAlgebra classes

   Ⓑ 60 students regardless of grade

   Ⓒ 60 students in the library

   Ⓓ 60 students in 7th grade

   Ⓔ 60 students in each grade

2. Researcher Bob is researching trees in a tree farm. Bob finds there are five different types of trees in this farm, and wants to know how many trees there are of each type. However, he also thinks it would take too long counting them all, one-by-one. Which of the following is the best way of knowing how many trees there are of each type?

   Ⓐ Visit the most dry area, and counts 1000 trees

   Ⓑ Visit one random area, and counts 1000 trees

   Ⓒ Visit the most central area, and counts 1000 trees

   Ⓓ Visit 10 different random areas, and counts 100 trees per area

   Ⓔ Visit the closest area, and counts 1000 trees

3. There are 9 white balls and 5 black balls, and one ball is randomly chosen. If a is the probability of choosing a white ball, b is the probability of choosing a black ball, and c is the probability of choosing a red ball, what is the value of $a + b + c$?

   Ⓐ $\dfrac{23}{14}$ Ⓑ $\dfrac{45}{196}$ Ⓒ $\dfrac{5}{7}$

   Ⓓ 1   Ⓔ $\dfrac{1}{14}$

4. Below is the number of words of five randomly selected pages of a 180 page book. What is the average number of words per page of this book?

   | 11, | 12, | 13, | 11, | 13 |

   Ⓐ 13   Ⓑ 12   Ⓒ 60

   Ⓓ 61   Ⓔ 11

5. Which of the following is the most appropriate statement?

Ⓐ The probability of 0.2 is less likely to occur that that of a probability of 0.7

Ⓑ All of other choices

Ⓒ The probability of an event is a number between 0 and 1 including 0 and 1

Ⓓ The probability of 0 represents an event never happens

Ⓔ The probability of 1 represents an event always happens

6. Which of the following probabilities is the least likely to occur?

Ⓐ 1   Ⓑ 0.9   Ⓒ 0.2

Ⓓ $\dfrac{5}{7}$ Ⓔ $\dfrac{1}{2}$

7. There are 39 white balls, 3 red balls, and 18 black balls in a large bag. What is the probability, in percent, to select a white ball if one ball is selected randomly?

Ⓐ 6.5%   Ⓑ 55%   Ⓒ 60%

Ⓓ 0.65%   Ⓔ 65%

8. Megan is rolling a fair die. After 54 rolls, how many times would she expect that 5 would be rolled?

Ⓐ 45   Ⓑ 36   Ⓒ 27

Ⓓ 18   Ⓔ 9

Answer Keys (Scan item or test QR code to get detailed solution steps)

Skill Practice (7.RP.A.1)
    1. B   **2.** 2   **3.** 15   **4.** 7

Skill Practice (7.RP.A.2)
    1. E   **2.** $P = 7d$ **3.** 297   **4.** 24

Skill Practice (7.RP.A.2.A)
    1. B   **2.** B   **3.** 8   **4.** A

Skill Practice (7.RP.A.2.B)
    1. 4   **2.** 9   **3.** $1.89   **4.** $216

Skill Practice (7.RP.A.2.C)
    1. E   **2.** $d = 3t$ **3.** $C = 11n$ **4.** 40

Skill Practice (7.RP.A.2.D)
    1. 2   **2.** B   **3.** 4.5 hours   **4.** E

Skill Practice (7.RP.A.3)
    1. 10%   **2.** E   **3.** 7800   **4.** $16.28

Comprehensive Exam (7.RP)
    1. E   **2.** D   **3.** A   **4.** B
    5. C   **6.** D   **7.** E   **8.** E
    9. A   **10.** A   **11.** B   **12.** B
    13. A   **14.** D   **15.** C   **16.** E

Skill Practice (7.NS.A.1)
    1. B   **2.** -29 **3.** $-$$1.64   **4.** -0.3 millions

Skill Practice (7.NS.A.1.A)
    1. -4 **2.** 0 dollars   **3.** $-16,\ 16$ **4.** 5000 feet

Skill Practice (7.NS.A.1.B)
    1. B   **2.** -29 **3.** 41 feet   **4.** A

Skill Practice (7.NS.A.1.C)
    1. $148.3^o F$ **2.** $67^o C$ **3.** $-61^o F$ **4.** 218.2

Skill Practice (7.NS.A.1.D)
    1. $20\frac{1}{11}$ **2.** $13\frac{6}{7}$ **3.** $19.53   **4.** 12.11

Skill Practice (7.NS.A.2)
    1. 145   **2.** 3.227   **3.** 30   **4.** $12.5^o F$

Skill Practice (7.NS.A.2.A)
    1. -6.5 **2.** -66.6 **3.** 4.48   **4.** $130.62

Skill Practice (7.NS.A.2.B)

1. 5 **2.** B **3.** 15.5 **4.** 7

Skill Practice (7.NS.A.2.C)

1. $\dfrac{5}{21}$ **2.** $-\dfrac{12}{125}$ **3.** C **4.** $\dfrac{3}{4}$

Skill Practice (7.NS.A.2.D)

1. C **2.** 1.37 **3.** 7.67 **4.** 0.32

Skill Practice (7.NS.A.3)

1. $6.63 **2.** $3.97 **3.** 72 **4.** $\dfrac{527}{35}$

Comprehensive Exam (7.NS)

1. B **2.** D **3.** A **4.** B
5. C **6.** D **7.** E **8.** E
9. A **10.** A **11.** A **12.** B
13. A **14.** D **15.** C **16.** E

Skill Practice (7.EE.A.1)

1. $-3x + 5$ **2.** $-7(5x + 9y)$ **3.** $-2x + 11$ **4.** -66

Skill Practice (7.EE.A.2)

1. $16w$ **2.** $x + 0.16x$ **3.** B **4.** D

Skill Practice (7.EE.B.3)

1. -18 **2.** 28 **3.** 226 **4.** 83.25%

Skill Practice (7.EE.B.4)

1. $8n \geq 25$ **2.** $34 + w \leq 50$ **3.** B **4.** 58

Skill Practice (7.EE.B.4.A)

1. 16 **2.** $t = -72$ **3.** $40 **4.** 8

Skill Practice (7.EE.B.4.B)

1. $x \leq -5$ **2.** $x > 8$ **3.** D **4.** 3

Comprehensive Exam (7.EE)

1. E **2.** D **3.** A **4.** B
5. C **6.** C **7.** E **8.** E
9. A **10.** A **11.** B **12.** B

Skill Practice (7.G.A.1)

1. 7 **2.** 105 **3.** 10 **4.** 12 ft by 7.5 ft

Skill Practice (7.G.A.2)

1. E **2.** 110° **3.** B **4.** D

Skill Practice (7.G.A.3)

1. Rectangle **2.** D **3.** Triangular Prism **4.** A

Skill Practice (7.G.B.4)

 1. 254.34   **2.** 64π **3.** 9.42   **4.** 9π

Skill Practice (7.G.B.5)

 1. 4   **2.** 100   **3.** 69   **4.** 136°

Skill Practice (7.G.B.6)

 1. 210   **2.** 336   **3.** 769.3   **4.** $61.34

Comprehensive Exam (7.G)

 1. E   **2.** E   **3.** A   **4.** B
 5. C   **6.** C   **7.** E   **8.** E

Skill Practice (7.SP.A.1)

 1. B   **2.** C   **3.** E   **4.** A

Skill Practice (7.SP.A.2)

 1. 2040   **2.** A   **3.** E   **4.** B

Skill Practice (7.SP.B.3)

 1. E   **2.** C   **3.** E   **4.** A

Skill Practice (7.SP.B.4)

 1. B   **2.** A   **3.** a, c   **4.** D

Skill Practice (7.SP.C.5)

 1. B   **2.** C   **3.** red   **4.** 57

Skill Practice (7.SP.C.6)

 1. D   **2.** 5   **3.** 33   **4.** 28%

Skill Practice (7.SP.C.7)

 1. $650   **2.** C   **3.** $\dfrac{13}{19}$ **4.** $\dfrac{1}{2}$

Skill Practice (7.SP.C.7.A)

 1. C   **2.** $\dfrac{3}{11}$ **3.** $\dfrac{55}{72}$ **4.** $\dfrac{7}{16}$

Skill Practice (7.SP.C.7.B)

 1. C   **2.** C   **3.** $\dfrac{15}{19}$ **4.** $\dfrac{3}{5}$

Skill Practice (7.SP.C.8)

 1. $\dfrac{7}{12}$ **2.** $\dfrac{5}{6}$ **3.** $\dfrac{1}{8}$ **4.** $\dfrac{23}{24}$

Skill Practice (7.SP.C.8.A)

 1. $\dfrac{1}{12}$ **2.** E   **3.** $\dfrac{1}{6}$ **4.** $\dfrac{1}{6}$

Skill Practice (7.SP.C.8.B)

1. B **2.** 36 **3.** 12 **4.** $\dfrac{1}{48}$

Skill Practice (7.SP.C.8.C)

1. B **2.** 16 **3.** C **4.** $\dfrac{3}{91}$

Comprehensive Exam (7.SP)

1. E **2.** D **3.** A **4.** B

5. B **6.** C **7.** E **8.** E

www.ingramcontent.com/pod-product-compliance
Lightning Source LLC
Chambersburg PA
CBHW051421200326

41520CB00023B/7320